LENSES!
TAKE A CLOSER LOOK

Can you see well?

Place the book open on a table and sit three big steps away from it on a chair. Close one eye and look at each picture of Lupi, the little creature on the top of the page. Which way is he pointing with his arm or his leg? Try it with the other eye. If you can't see Lupi well, discuss it with your parents.

Table of Contents

① INUITTI, Eskimo girl with snow goggles
② LUPI, helper, discoverer, and game-maker
③ SHERLOCK HOLMES, detective
④ REINER WOLF, forester
⑤ DR. SEBASTIAN SHARP, eye doctor
⑥ ANTON van LEEUWENHOEK, microbiologist
⑦ GALILEO GALILEI, astronomer
⑧ KATHERINE
⑨ DENNIS
⑩ NICK

LENSES!

TAKE A CLOSER LOOK

Siegfried Aust ▪ illustrated by Helge Nyncke

Lerner Publications Company · Minneapolis

Take a Look at Your Eyes

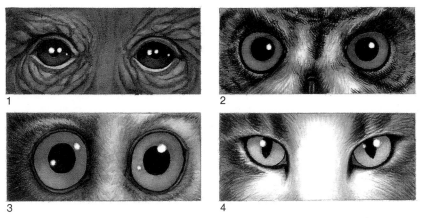

1

2

3

4

Look in the mirror and examine your eyes. How do your eyebrows and eyelashes look? What about the **scleras** (white sections), **irises** (colored disks), and **pupils** (black circles) of your eyes?

Katherine has to use detective work to identify the crook. She can only see the crook's eyes. Try playing the crook-hunting game with friends. Consider the color, size, and shape of the crook's eyes.

Who do the eyes belong to—an elephant, cat, owl, monkey, kangaroo, or orangutan? (Solution on page 31.)

When you look straight ahead, you have a fixed **field of vision**. If you stood inside a clock and looked at the number 6, your eyes would take in the scene between the 3 and the 9. A bird could see everything between the 1 and the 11. An insect's field of vision extends almost all the way around.

4

Seeing and Perceiving

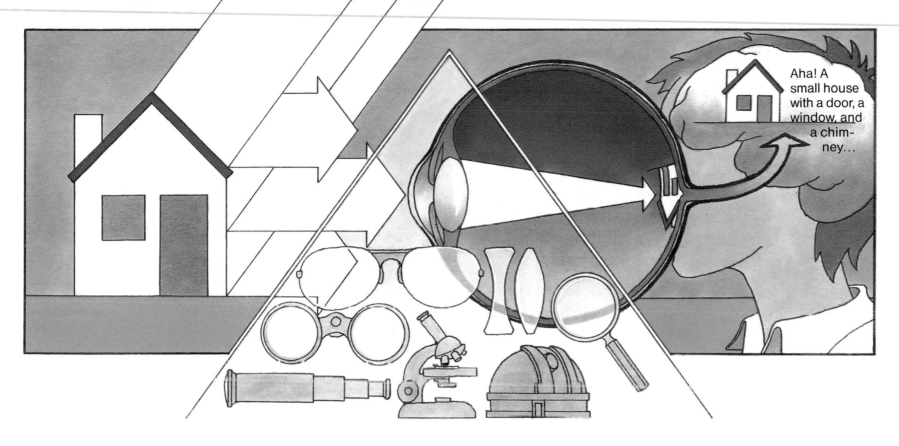

In order for us to see, we must have a light source (daylight from the sun or artificial light from a lamp, for instance). Because some light sources are brighter than others, the same object may look different when seen by moonlight, under a yellow streetlight, and at sunset.

When light rays strike an object, they reflect off its surface, and an image of the object is then sent through the air with the light rays.

Next, a **lens** takes the "light ray image" and **focuses** it, or makes it sharp. Our eyes have lenses that focus the images that pass through to our brains. In order to enlarge images of distant or tiny objects, we need additional lenses such as magnifying glasses, microscopes, telescopes, and eyeglasses.

The light rays must come to a precise point on a surface where the image will be fixed. In our eyes, this happens on the **retina**, a surface at the back of our eye. The retina is composed of millions of light-sensitive cells that enable us to distinguish colors and images. If these cells don't work correctly, they pass on incorrect messages to the brain. This happens, for example, when someone is **color-blind**.

Finally, our brains receive the light rays and we understand what is in front of us. There are many rays that our eyes cannot see, however, such as heat rays. Therefore, we cannot see whether an object is hot or cold.

The brain can see pictures that we produce inside our own heads, though. This happens when we dream or when—even with our eyes open—we "daydream," or visualize something that we're not looking at.

Seeing Clearly and Recognizing Colors

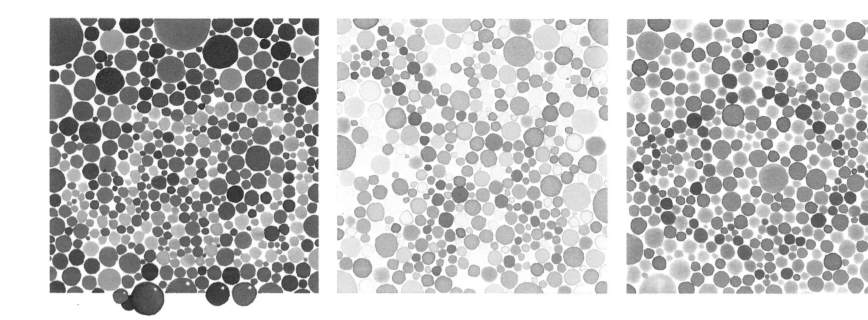

In each picture above, animals are hidden among the dots. Someone who is color-blind might not be able to find the animals. People who want to get a driver's license must not only see well, but must also be able to recognize colors. Otherwise, they might run a red light, thinking it's green. People also use colors as symbols. Red frequently means "attention!" or "danger!"

In Front and Behind—Depth Perception

The pirates are throwing dice to divide up the loot. Every time a 6 is rolled, each pirate grabs at the coins in the middle of the table. But one pirate comes out badly every time. Do you know why?

The answer is simple. The pirate on the left sees with only one eye. You can do a test to understand how hard it is to judge distances with only one eye.

Lay this book about one arm's length away from you on the table. Shut one eye. Now, with one finger, quickly touch the coin at the bottom of the page. Did you hit it exactly?

When you see with both eyes, each eye views an object from a slightly different angle. This gives you three-dimensional vision and allows you to judge distance and thickness accurately. With one eye closed, a person has less **depth perception**.

7

The First Lenses Are Developed

In ancient Rome, the writer Seneca discovered that images became enlarged when seen through a glass ball filled with water. Try it yourself. Fill a glass bottle with water and hold it in front of your face.

During the Middle Ages in Italy, people cut glass lenses that could make things appear bigger or smaller. There are two kinds of lenses. **Convex lenses** are thicker in the middle than at the edges. When rays of light pass through this type of lens, they are bent inwards and meet at a focus point behind the lens. Convex lenses are also called "condensers."

Lenses that are thicker at the edges than at the center are called **concave lenses**. When light rays pass through these lenses, they diverge, or spread outwards. For this reason concave lenses are also called "diverging lenses."

Convex and concave lenses are used in eyeglasses, microscopes, cameras, and other devices to change images in many different ways.

concave lens

convex lens

focus point

The 700-Year History of Lenses

Around 1270, people began to make lenses out of rock crystals called beryl. Roger Bacon made the first eyeglasses in Europe around 1287. The first picture of a person wearing glasses dates from 1352. The painting shows a monk reading with eyeglasses. In Germany, the artist Konrad von Soest painted a self-portrait in which he holds a pair of glasses in a wooden frame. The painting dates from 1403.

People used to hold their eyeglasses in front of their eyes with handles or fasten them with clips, rings, or ribbons. Modern eyeglasses fit over your ears and rest on the bridge of your nose. The first eyeglass frames were usually fairly heavy. They were made out of metal, wood, tortoiseshell, or even ivory. Modern eyeglasses are generally made out of light metal or plastic and are much more comfortable to wear.

In earlier times, some people wore **monocles**. A monocle is a glass for only one eye. Some people held their monocle with a handle. Others held the glasses in place by pinching them tightly in their eyesockets. Try this with a half-dollar.

A Visit to the Eye Doctor

① Katherine can see well at a distance. When she wants to read, however, she can barely make out the writing. "Katherine is **farsighted**," the eye doctor explains. She needs a pair of glasses for looking at things close up. The doctor tries out lenses of different thicknesses until he finds the ones that help Katherine read the small letters on the chart. The doctor writes down the strength of these lenses on a prescription.

② Uncle Otto is **nearsighted**. He can read well without glasses. However, he only recognizes his friends when they are standing very close to him. Uncle Otto's glasses help him see at a distance. Eyeglasses for nearsightedness, farsightedness, and other vision problems have differently-shaped lenses.

③ Some people cannot see well close up or from far away. They need two pairs of glasses. Or they might wear **bifocals**—glasses with two prescriptions in one lens. The part for reading is on the bottom of the lens; the part for distance is on top.

Shopping at the Optician's Store

The optician takes Katherine's prescription and orders the lenses from a laboratory. She does not have the lenses in stock the way a supermarket has food. Everyone needs his or her special glasses.

Katherine chooses her eyeglass frames from a large selection. "There's a style here for everyone," says the optician. "Find the pair that makes you look your best." Someday, Katherine might want **contact lenses**—prescription lenses made of thin pieces of plastic that fit directly on a person's eyes.

At the laboratory, the lenses are made to the right thickness and shape to fit Katherine's prescription and eyeglass frames. The optician makes sure the eyeglasses fit Katherine's face comfortably. She gives Katherine a sturdy storage case and a cloth for cleaning her glasses and wishes her "good seeing."

11

Optical Illusions

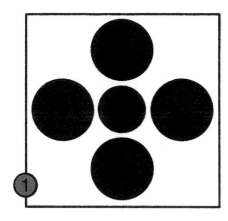

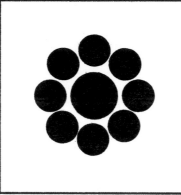

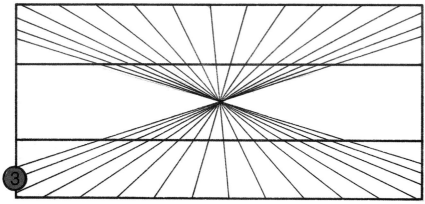

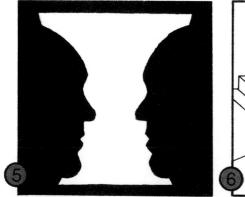

Sometimes our eyes can play tricks on us. Look at pictures 1 through 7 carefully and answer these questions:

1 Which of the two circles in the middle is larger?

2 Both Lupis are the same size, right?

3 Are the horizontal lines straight or curved?

4 Look at the picture on the left for one minute. Now look at the one on the right. Has the color of Katherine's face changed?

5 What do you see?

6 How can one person go up the steps, and another go down the steps at the same time?

7 Where is the thin stripe lighter? On the left or on the right?

(Solutions on page 31.)

When People Cannot See

You've probably seen a person who uses a white cane. This person is either blind or has severely damaged vision. A person can be born blind or can become blind through injury, infection, or illness. Blind people rely on their senses of hearing and touch to find out what's going on around them.

In earlier times, blindness was a terrible handicap that left people completely helpless. In modern times, however, there are many aids for the blind. With a white cane, the blind can "feel" their way along streets and hallways. In addition, the white cane is an identifying mark showing that a person is blind and might need assistance. Some blind people use a guide dog to lead them.

Nearly 200 years ago, the Frenchmen Charles Barbier and Louis Braille developed a special alphabet system for the blind. This system, called **Braille**, is based on six dots. Each different combination of raised dots stands for a different letter of the alphabet.

To find out what it is like to use Braille, draw dots on heavy paper for each letter of the alphabet as shown. Now use a pencil or pointed object to press the dots up from underneath the paper. You can run your fingers over the bumps to "read" the letters on the paper. Special typewriters also let blind people write books and letters in Braille.

A B C D E F G H I J

K L M N O P Q R S T

U V W X Y Z

. , ! ( ) ? :

Observed through the Magnifying Glass

Can you see the tiny writing on these postage stamps? If you can't, try using a **magnifying glass**.

Magnifying glasses use a convex lens to make images look bigger than they really are. Magnifying glasses are also called "burning glasses" because when sunlight passes through the convex lens, the rays gather at one point and can become hot enough to burn a flammable object.

Magnifying glasses were first used in Europe during the Middle Ages. For some types of work, our eyes are not adequate. Watchmakers, for example, must build and repair the tiny machinery and gears within watches. They may hold a magnifying glass up to their eyes to help them perform this delicate work.

Look through a magnifying glass and you can see many things more precisely than you can with your naked eye. Explore your environment! How does a snowflake, a grain of salt, or a strand of hair look under the magnifying glass?

When sunlight hits the magnifying glass at noon, it starts a fire that shoots off the cannon.

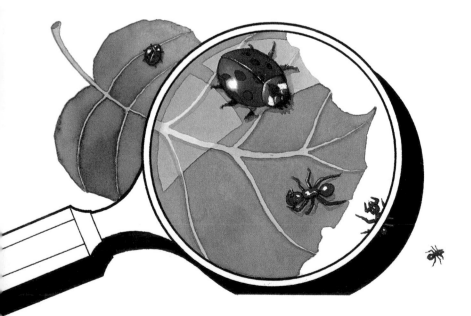

The most powerful magnifying glasses can enlarge an image clearly to 20 times its original size.

The 400-Year History of the Microscope

In 1590 the Dutch Janssen brothers built the first **microscope**. A microscope is more powerful than a magnifying glass because it has two lenses. The **objective lens** on the bottom works just like a magnifying glass to gather light and enlarge an image. The **ocular lens** on the top of the microscope further magnifies the image created by the objective lens.

In 1676, the Dutchman Anton van Leeuwenhoek used the microscope to study the many microorganisms that populate our world.

In 1900, in Germany, Ernst Abbé developed a microscope that magnified images up to 2,000 times. Thus scientists could study the tiniest living things, such as bacteria and viruses that cause dangerous diseases. The power of microscopes has continued to improve. Microscopes are valuable tools in medicine and the study of nature.

You Can Learn Microscopy

onion skin

waterweed

DIA-FIX

We observe the object through the **ocular lens** of the microscope.

The image is sharply focused with the **coarse adjustment knob**.

Objective lenses with varying magnifications are chosen from a **revolving nosepiece**.

The glass slide is placed on the **stage**. It is secured with a cover glass.

A mirror and a lamp illuminate the object on the slide.

Dennis and Katherine have dripped water from a dropper onto a glass slide and placed the slide under the microscope. They are amazed at how much there is to see when the light shines through the water. They see tiny living things that cannot be seen with the naked eye.

Dennis and Katherine also examine such things as a cross-section of an onion skin and a piece of waterweed under the microscope.

Like a slide projector, the microscope must have a powerful lamp to illuminate the object we want to see. The lamp shines onto a mirror that reflects light through the object, the magnifying lenses, and into the eyepiece.

The Electron Microscope Shows the Invisible

Under the magnifying glass, we can see that television and newspaper pictures are made up of many small dots. Without magnification, the distance between the dots is invisible and the pictures appear to be made up of solid colors.

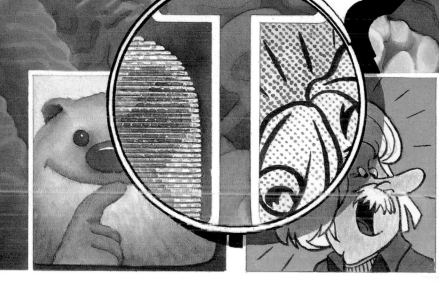

The standard microscope works with light rays. With this microscope, we can see things that are larger than 1/1000 millimeters. Everything smaller remains invisible. If you want to see things that are smaller than 1/1000 millimeters, such as atoms, you need a microscope that works with electron rays. The first electron microscope was built over 50 years ago. Modern scientists now have electron microscopes that can magnify things by more than 1,000,000 times.

What could the strange picture in the background be? (Solution on page 31.)

Far Away and Yet Clear

Whom have the sailors discovered?

With the microscope, you can investigate the smallest things in your everyday environment. But if you want to uncover the secrets of the faraway, you must make them seem close to you. This can be done with a **telescope**. This device could really be called a near-scope, since through the telescope distant things are brought right before your eyes.

④

The History of the Telescope

In 1608 in Holland, Hans Lippershey ① made a telescope by placing lenses at each end of a tube. With it he could see the weathervane on the church steeple quite clearly. In 1609, the Italian scholar Galileo Galilei ② constructed a more powerful telescope. With his invention he could see craters on the moon, the moons of Jupiter, and the rings of Saturn. In 1612 in Germany, Simon Marius ③ developed a more powerful telescope and identified the group of stars, or constellation, called Andromeda.

Early telescopes were **refracting telescopes**. They had an objective lens that gathered light at one end of the tube and ocular lenses that magnified the image at the other end. But there were many problems with early telescopes. Images appeared with multicolored ridges. In 1611, Johannes Kepler ④ built a telescope that didn't show rainbow colors and ridges.

Look to the Stars—the Reflecting Telescope

Astronomic observation site, Stonehenge, southern England, ca. 2000 B.C.

Refracting telescope, Paris, about 1680

Reflecting telescope, England, 1790

People began observing the stars thousands of years ago. In some countries, their observation posts still stand. Stonehenge in the south of England is believed to be such an observation post. Without telescopes, people could only watch the movement of the stars in the sky.

With the invention of the telescope it became possible to study the stars more closely. People built large observatories and made larger and larger telescopes. **Reflecting telescopes** use mirrors instead of objective lenses to gather light. Reflecting telescopes are larger and can gather more light than refracting telescopes and thus allow people to see objects that are further away and dimmer. The Hale Telescope was set up at Palomar Observatory in California in 1948. This telescope has a 16.5-foot (5-meter) wide reflecting mirror. Large telescopes are usually set up on high mountains so that there will be less dust and smog to cloud our view of outerspace. The Hubble Space Telescope, however, actually travels *in* outerspace. The telescope is housed inside a spacecraft. As it orbits the earth, it allows scientists to view the far regions of the universe.

Look into Space—the Radio Telescope

Reflecting telescope,
Mount Palomar,
California, 1948

Radio telescope, Australia, about 1980

Amateur astronomers usually
use reflecting telescopes.

Just as the electron miscroscope developed from the standard microscope, the **radio telescope** developed from the reflecting telescope. Like the microscope, the reflecting telescope uses light rays to illuminate distant objects. But if one wants to see farther into space, special technology is needed. The radio telescope gathers radio waves instead of light waves and looks like a giant antenna. With this telescope, we can *listen* to the deepest regions of space—much farther than we can see.

A Visit to the Observatory

Katherine, Dennis, and Nick are visiting an observatory. They learn that the great astronomic instruments are no longer refracting telescopes. They are reflecting telescopes with enormous mirrors. These telescopes stand in large observatories. The largest telescopes enable people to look three sextillion miles into space. We can hardly imagine this distance. A sextillion is a number with 21 zeroes. The reflecting telescopes stand in giant domes, weigh hundreds of tons, and are moved into different positions by electric motors.

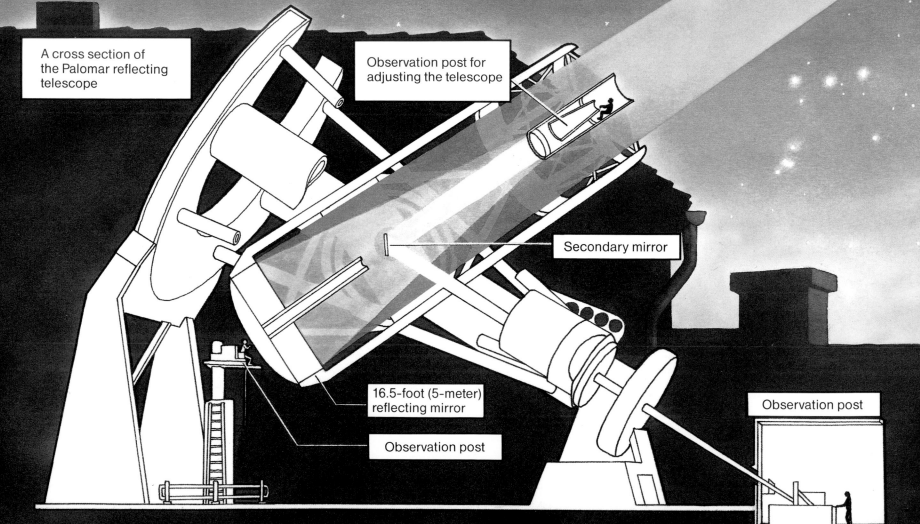

A cross section of the Palomar reflecting telescope

Observation post for adjusting the telescope

Secondary mirror

16.5-foot (5-meter) reflecting mirror

Observation post

Observation post

The Wide, Starry Sky above Us

On a clear night you can observe the stars. If you compare the position of the stars over several hours, you will observe that most stars "wander." Only a few stars appear to stay fixed in place. In the night sky above, you see the moon and a cluster of brightly shining stars called Pleiades above Lupi.

The stars of the constellation Orion shine above the chimney in the middle.

Nature Observed Precisely

Dennis, Katherine, and Nick are watching different types of birds in a nature preserve and are looking up the correct names in a guidebook. Dennis watches the birds through the telescope. Katherine can see many details with her **binoculars**, or field glasses. Nick would rather look at frogs, since without a visual aid he cannot see the birds well! The first telescopes were held by hand or set up on a stand or tripod. With a telescope or binoculars, our field of vision is quite small. It is not easy to get all of the desired object in view. Try to see a particular star with binoculars or a telescope.

Seeing Far Away with a Telescope and Binoculars

People often use binoculars or opera glasses to watch performances, sporting events, or wildlife. Both are double telescopes—they have an eyepiece and a lens for each eye. At first people made binoculars simply by holding two single telescopes next to each other. These were very long, awkward, and heavy.

In 1900, Ernst Abbé, at the Carl Zeiss optical factory in Jena, Germany, built a pair of binoculars in which the light was reflected through **prisms**. Prisms are triangular glass pieces that function like mirrors to reflect light through the binoculars. Prism binoculars provide greater magnification and give a wider field of view than do binoculars without prisms. They are also smaller and easier to handle.

If you look at the cross section of the binoculars in the picture, you can follow the path of the light from the objective lens to your eye.

If you asked a salesperson "How powerfully do binoculars magnify?" "How large will my field of view be?" and "How well can I see something at night?" you would be amazed at how many different types of binoculars there are.

Silly Question:

How do you catch an elephant with binoculars, a matchbox, and tweezers?

(Solution on page 31.)

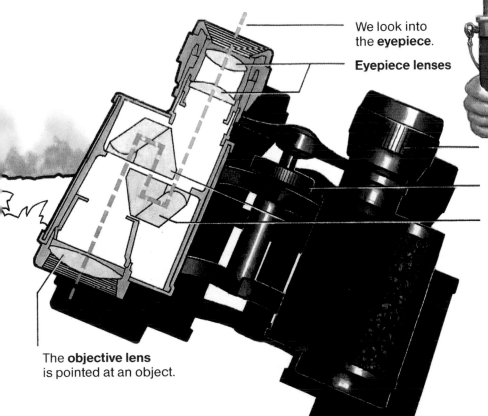

We look into
the **eyepiece**.

Eyepiece lenses

Focusing device for the individual tube

Focusing wheel for focusing both lenses together

Prisms reflect the light.

The **objective lens**
is pointed at an object.

Let's Build a Periscope

Nick has made a device for peeking around corners and over walls. To build a periscope, you need: Two mirrors, 2 x 3 inches (5 x 7.5 cm); a sheet of black cardboard, 8.5 x 11 inches (21 x 27.5 cm); a toilet paper roll; tape; a ruler; a colored pencil; and a cutting knife.

Draw four lines from the top to the bottom of the cardboard. The lines should be 2 inches (5 cm) apart. Now draw one line across the cardboard 2 inches (5 cm) from the top and another 2 inches (5 cm) from the bottom. Cut out the eyeholes and the sections marked with an X in the picture. Fold the cardboard along the vertical lines to form a long, thin box. Tape the box closed and tape one mirror to each end. Make an eyepiece with a section of the toilet paper roll.

2 inches

7.5 inches

2 inches

2 inches

The Many Colors in the Kaleidoscope

To make a kaleidoscope you will need: heavy black cardboard; two mirrors, 2.5 x 8 inches (6 x 20 cm); cellophane or a piece of thin plastic; multicolored scraps of paper; scissors; tape.

Cut a piece of black cardboard 2.5 x 8 inches (6 x 20 cm) ①. Make a three-sided tube by taping the two mirrors and the cardboard together as shown. The mirrors should face inward so that the inside of the tube has two reflective sides and one black side ②. Cut out a cardboard triangle with a peephole as shown ③ and tape the triangle to one end of the tube. Make a triangular pouch out of cellophane or plastic as shown in the picture ④ and fill it with scraps of colored paper. Tape the pouch in place at the other end of the tube. When you look through the peephole you can see colorful designs that change as you turn the kaleidoscope.

③

④

②

①

Dangers for the Eyes

Our eyelashes and eyelids protect our eyes from dust, sand, and other things that might harm them. If something does get into your eye, it can often be flushed out with water or tears. But sometimes, you must go to the eye doctor if something harmful gets into your eyes.

Nick and Dennis arranged a bike race. Dennis had a strong lead when suddenly, Nick overtook him easily. What in the world was wrong with Dennis? He'd gotten an insect in his eye!

Yes, this time Dennis's eyelashes didn't work as guards for his eye. During the fast ride, the insect hit Dennis's eye so suddenly that the eye couldn't shut quickly enough to protect itself. Normally, your eye will automatically blink shut to protect itself from a harmful object.

The eyes are especially sensitive to glaring light, wind, or irritants such as chlorine in pool water. For divers, welders, skiers, and so on, there are various types of glasses to protect the eyes. In the picture below, in fact, there are several kinds of protective glasses. Can you find the one that doesn't belong?

28

The Eyes Are Examined

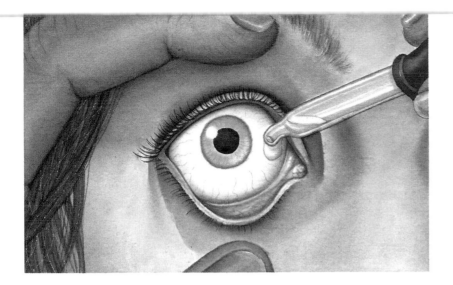

By having regular eye examinations, wearing proper eyeglasses and protective glasses, and keeping dangerous objects away from your eyes, you can help ensure that you will have healthy eyes for a whole lifetime.

The eye doctor is examining Katherine's eyes. He shines a light into her eyes and inspects the retinas. To make the pupils larger, he put a couple of drops of fluid in Katherine's eyes before he examined them. Katherine blinked her eyelids when the doctor put his instruments near her eyes. The doctor explained to her, "Our eyes are very sensitive and must be protected. They can also protect themselves. That's why we shut our eyes when we sense danger. Our eyes might begin to tear to flush away something that is irritating them. Your eyesight can be damaged if you sit too close to the TV screen or read in poor light for a long time."

The eye doctor also treats eye diseases and eye injuries. It is important to have a doctor give you regular eye examinations to check for disease, nearsightedness, farsightedness, and other vision problems. The eye doctor will also check your eyeglasses or contact lenses and write a prescription for new lenses if your vision has changed since your last visit.

Katherine observed everything the eye doctor did with interest. She reported to Dennis and Nick, "It didn't hurt, but sometimes it was uncomfortable."

All about Eyes

There are many interesting and noteworthy things to discover about eyes in the animal world, too. You'll find some examples here in the form of riddles—you can look for the animals that match the answers.

Q. What animal has the biggest eyes?
A. The giant squid has eyes as big as dinner plates.

Q. What animal has the most eyes?
A. Insects. Dragonflies have faceted eyes with over 20,000 individual lenses.

Q. What animal has the sharpest eyes?
A. The sea eagle, which can recognize fish in the water from high above.

Q. What animal has eyes that don't see?
A. The blind cave salamander.

Q. What animal can see in almost total darkness?
A. The owl.

Q. What animal's eyes shine in the dark?
A. The cat's eyes shine when light hits them.

Q. What animal can see in two different directions at the same time?
A. The chameleon can move each eye separately.

Q. What animal has eyes only on one side of its face?
A. The flounder. The side without eyes lies on the sea floor.

Q. What animal wears glasses that it can't see through?
A. The cobra has a pattern on its back that looks like eyeglasses.

Q. What animal likes to steal people's glasses even though it sees well?
A. The magpie likes to collect shiny, reflective objects.

Q. What animal can move its eyes in and out?
A. The snail has eyes on stalks.

Q. What animal has false eyes?
A. The calico moth has false eyes on its wings to frighten enemies. It also has real eyes.

Solutions to Puzzles

page 4: 1. Orangutan
2. Owl
3. Monkey
4. Cat

page 12: 1. The circles are the same size.
2. The Lupis are the same size.
3. The horizontal lines are straight.
4 Katherine's face looks reddish.
5. It could be a vase or two faces.
6. It is not possible. The picture is playing a trick on your eyes.
7. It is the same shade everywhere.

page 17: The picture shows the surface of the tongue of a housefly.

page 25: Hold the binoculars the other way around and look at the elephant. It looks quite small now. Grab it with the tweezers and put it in the matchbox.

31

Author Siegfried Aust loves both technology and writing for children. Aust has combined his interests in the Fun with Technology series. He is a teacher who has written many books for children, including *Flight! Free as a Bird.*

Illustrator Helge Nyncke lives in Muhlheim-am-Main, West Germany, with his wife and children. Nyncke studied at the University for Design in Offenbach and has illustrated many children's books, including textbooks, stories, and poetry.

This edition first published 1991 by Lerner Publications Company. All English language rights reserved.

Original edition copyright © 1987 by Verlag Carl Ueberreuter, Vienna, under the title *Schau her, ich seh noch mehr: Von Lupen, Brillen und Fernrohren.* Translation copyright © 1991 by Lerner Publications Company. Additional text and illustrations copyright © 1991 by Lerner Publications.

Translated from the German by Amy Gelman.

Library of Congress Cataloging-in-Publication Data

Aust, Siegfried.
  [Schau her, ich seh noch mehr. English]
  Lenses! take a closer look / Siegfried Aust ; illustrated by Helge Nyncke.
    p. cm.
  Translation of: Schau her, ich seh noch mehr.
  Summary: Explains all about lenses, magnifying glasses, telescopes, microscopes, eyeglasses, vision, and eye care.
  ISBN 0-8225-2151-2 (lib. bdg.)
  1. Lenses—Juvenile literature. [1. Lenses.] I. Nyncke, Helge, ill. II. Title.
QC385.A9713 1991
681′.4—dc20
                                90-39260
                                  CIP
                                  AC

Manufactured in the United States of America

2 3 4 5 6 7 8 9 10 99 98 97 96 95 94 93 92